The Future Of Medicine: Healing Yourself Regenerative Medicine

Part One

William A. Haseltine PhD

ACCESS Health Press

Recent Books by William A. Haseltine

Affordable Excellence: the Singapore Healthcare Story; William A Haseltine (2013)

Improving the Health of Mother and Child: Solutions from India; Priya Anant, Prabal Vikram Singh, Sofi Bergkvist, William A. Haseltine & Anita George (2014)

Modern Aging: A Practical Guide for Developers, Entrepreneurs, and Startups in the Silver Market; Edited by Sofia Widén, Stephanie Treschow, and William A. Haseltine (2015)

Aging with Dignity: Innovation and Challenge is Sweden-The Voice of Care Professionals; Sofia Widen and William A. Haseltine (2017)

Every Second Counts: Saving Two Million Lives. India's Emergency response System.The EMRI Story; William A Haseltine (2017)

Voices in Dementia Care; Anna Dirksen and William A Haseltine (2018)

Aging Well; Jean Galiana and William A. Haseltine (2019)

World Class. Adversity, Transformation and Success and NYU Langone Health; William A. Haseltine (2019)

Science as a Superpower: My Lifelong Fight Against Disease And The Heroes Who Made It Possible; William A. Haseltine (2021)

Living ebooks

A Family Guide to Covid: Questions and Answers for Parents, Grandparents and Children; William A. Haseltine (2020)

A Covid Back To School Guide: Questions and Answers for Parents and Students; William A. Haseltine (2020)

Welcome to *The Future of Medicine: Healing Yourself, Regenerative Medicine | Part One!*

Please find the contents of this lecture I gave my United States team of ACCESS Health International in August 2023. ACCESS Health International is a nonprofit think tank, advisory group, and implementation partner dedicated to improving access to high-quality, affordable healthcare globally.

This book covers some key topics in regenerative medicine. It serves as part one, and more parts to follow. You can find more information about regenerative medicine and download the future parts and updates to the regenerative medicine series by visiting

www.williamhaseltine.com/regenerative-medicine

Thank you for your interest.

Acknowledgments

I thank the ACCESS Health US team: Courtney Biggs, Koloman Rath, Amara Thomas, Griffin McCombs, Roberto Patarca, and Kim Hazel for their support in creating this book.

This work is supported by ACCESS Health International (www.accessh.org).

Dedication

As always, I am deeply grateful to my wife, Maria Eugenia; my children Mara, Alexander, Karina, Manuela, and Camila; and delightful grandchildren, Pedro Augustin, Enrique Matias, and Carlos Eduardo, for their loving support.

Contents

Introduction

An acorn, small and unassuming, rests on the forest floor. It's at the mercy of harsh winds, blazing heat, torrential downpours, and forest-foraging animals. Yet, it remains unchanged... for a time. Gradually, the acorn sprouts anew as its root stretches out and down, diligently digging into the soft soil.

The acorn transforms into a sturdy trunk that grows tall and strong, reaching skyward. Lush greenery with abundant leaves burst forth from the branches in a splendid grandeur. These spread out in all directions, forming a benevolent umbrella that shields the world below.

This former acorn-turned majestic tree thrives for decades, enduring all the seasons of the world until the time comes for a turnover. Tiny buds appear on the bare branches of the aged tree, which turn into fresh acorns that drop and settle on the forest floor, ready to start anew.

Intrinsic to life is the idea that every living thing undergoes renewal.

A complete cycle of life that relentlessly marches on. The acorn is back where it began, fresh and new. From seedling to sapling to tree, it's a poetic reminder that every end is, in fact, a new beginning. This is the chief wonder of life.

Life endures as mountains wear away, then rise again, adorned in a lush living cloak. This cycle of renewal surrounds us – in the rhythmic tides of the ocean that sort and shift debris, in the resolute erosion that sculpts towering peaks. Yet renewal is found in us, too, manifesting on a microscopic scale.

DNA is an immortal, unifying molecule – existing both now and forevermore throughout all life. The cycle of life persists, an age-old narrative that began in the past far beyond our imagination. Immortal and unrelenting, it continuously renews and regenerates itself. It is this fundamental concept that fuels regenerative medicine.

Intending to restore our bodies to normal, whether injured by trauma, damaged by disease, or worn by time, regenerative medicine aims to tap into the endless cycle of renewal and rejuvenation that characterizes life.

By studying our natural healing abilities, we are unlocking new ways to regenerate tissues, improve outcomes and enhance the quality of life for people across the world. Whether it involves cell, gene, or protein therapies, or cutting-edge biomechanical interventions, regenerative medicine is revolutionizing the healthcare landscape as we know it.

A Cornerstone of Biomedical Innovation: DNA Technologies

At the heart of regenerative medicine lies one of its most significant elements: DNA. DNA is both a blueprint and a building block for the human body. It contains the instructions needed to create and maintain our bodies and is made of smaller units that link to form the iconic image that we know as the DNA helix. The study of genetics focuses on DNA and has allowed us to understand fundamental processes that govern growth, development, and disease response.

Researchers and physicians have harnessed this knowledge to create innovative therapies targeting genetic defects and inherited diseases. Moreover, advances in DNA research have been instrumental in developing novel antibodies, which have proven effective in the fight against a wide range of infectious diseases and cancers.

DNA technologies have revolutionized the field of pharmaceuticals and drug discovery. This has led to a considerable surge in the demand for DNA-based drugs. According to industry experts, this demand is expected to continue rising and reach a staggering value of over [$113 million by 2029.](#) The market for DNA repair drugs is also growing significantly. It is expected to generate a worth of [20 million by 2026](#).

The transformative power of DNA technologies extends beyond the realm of pharmaceuticals and drug discovery. It forms the foundation upon which all regenerative medicine research and development is built.

The marriage of DNA technologies and regenerative medicine has ushered in a new era of personalized medicine, where an individual's genetic makeup serves as the basis for targeted therapy. This redefines how we treat diseases while offering the potential for much more precise, effective, and long-lasting treatments.

With this knowledge, we can anticipate a future where patients benefit from compelling, comprehensive, customized healthcare solutions catering to their unique circumstances. It propels us toward a future where personalized medicine becomes the norm.

The Next Frontier: Antibodies as Medicine

The realization that DNA could be applied to antibodies to create medicines was a pivotal discovery in medical history. It led to substantial changes in biomedical sciences and the pharmaceutical industry. In the past, the pharmaceutical industry primarily relied on small-molecule chemicals. However, nowadays, the production of high-value products is heavily influenced by new technologies.

Antibody therapies have emerged as a leading treatment for many diseases. They have skyrocketed in popularity over the past five years and are now the top-selling drugs. The FDA has approved over 100 antibody-based medications. Each year roughly 20% of all newly approved drugs are biologics.

The robust pipeline of antibody drug development holds immense potential for significant market revenue. Consequently, the field of antibody-based pharmaceuticals is witnessing rapid expansion. This is evident from the growing accessibility of numerous antibody-based treatments, which are gaining prominence.

Some of the most effective cancer treatments are antibody therapies such as Herceptin, Avastin, Rituxan, Erbitux, Gertrude, Opdivo, and Tecentriq. Antibody therapy extends its utility beyond cancer treatment. It is also instrumental in addressing various ailments like rheumatoid arthritis (Humira), Crohn's disease (Remicade), and macular degeneration (Lucentis). These medications are among the most commonly prescribed.

Antibody-based medications are highly effective due to their ability to precisely pinpoint and neutralize specific antigens, making them an effective form of therapy. These drugs mimic the body's innate immune response and can bind to antigens with extreme accuracy, preventing their harmful effects. By targeting specific antigens, antibody medicine can provide a more targeted and precise approach to treating diseases, ultimately contributing to better patient health outcomes. The potential of antibody medicine is vast and promising, as it offers a wide range of benefits and therapeutic applications. To understand more about the role of antibodies in infectious disease, read *Monoclonal Antibodies: The Once and Future Cure for Covid-19*. This book summarizes the current state of antibody technology and its application in controlling the Covid-19 pandemic.

Combatting Cancer

Cancer is a profoundly sensitive and crucial subject within the healthcare industry, and for a good reason. In 2015, around 90.5 million people were affected by the disease. Four years later, the number rose significantly to 114.1 million, with 10 million losing their lives due to the illness.

Cancer is an all-encompassing term that encapsulates over two hundred illnesses that can develop in any part of the human body. Like all things in the human body, cancer begins with the cells.

Specifically, cancer occurs when cells grow and divide uncontrollably. This rapid growth forms a mass known as a tumor. These tumors can either be malignant, meaning they can rapidly spread to other areas of the body and invade nearby tissues, or benign, meaning they are not cancerous and won't extend beyond their point of origin. Tumors can form anywhere in the body, and some cancers, like blood cancers, do not even need a tumor to wreak havoc on the body.

For many years, the field of biomedical research has been dedicated to the crucial objective of finding a cure for cancer, and over time, a multitude of noteworthy advances have been made in this pursuit.

One of the first significant cancer treatments dates back to 1891. It was a rudimentary form of immunotherapy developed by William B. Coley, thus earning it the moniker "Coley's Toxin" in later years. In this therapeutic approach, doctors inject patients with a mixture of bacteria to kill cancer cells. The FDA subsequently approved a "new drug" based on these principles in 1963.

By 1970, researchers used a modified version of these toxins to treat bone and soft-tissue sarcomas. These modified toxins were called mixed bacterial vaccines or MBVs. The lifetime of these was short-lived, though, with the discoveries of natural killer cells, monoclonal antibodies, and checkpoint inhibitors.

Natural killer cells, or NK cells, are a type of white blood cell that can recognize and kill cancer cells. At the time of the first bone marrow transplant to treat leukemia, doctors did not know they were using NK cells. Around 2000, NK cell-mediated immunotherapy was officially recognized as a safe and effective treatment. Since then, the development of NK therapies has focused on optimizing the source of NK cells and enhancing the cytotoxicity of the cells.

Monoclonal antibodies were hot on the heels of NK cell discoveries. The therapeutic application of monoclonal antibodies emerged in 1975. Monoclonal antibodies are a class of targeted agents that specifically recognize and bind to antigens on the surface of cancer cells. There are various types of these antibodies, but the most successful have been IgG monoclonal antibodies.

Over the last few decades, the FDA has approved over 100 monoclonal antibody treatments. The 2022 monoclonal antibody market topped $186.6 billion, with an expected value of $609 billion by 2032.

These antibodies are also employed in treating COVID-19, aiding the battle against the virus that changed the world in 2020.

A more recent breakthrough is checkpoint inhibitors. These therapies unleash the immune system's inherent potential by blocking molecules that act as brakes on immune cells. By releasing the brakes, T cells can identify and combat tumors. The first of these therapies, ipilimumab, was approved in 2011 for the treatment of melanoma.

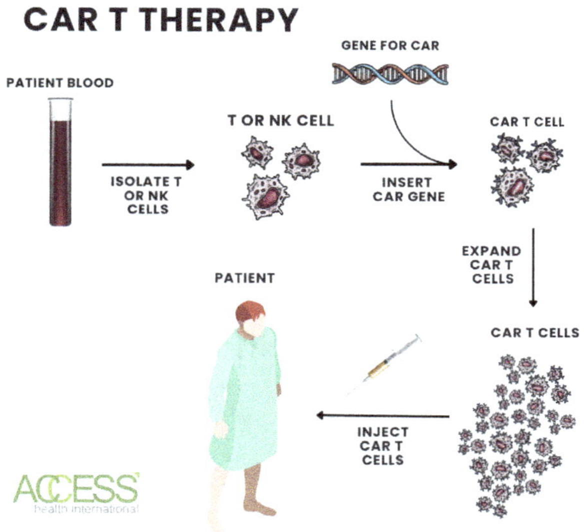

Figure 1. An infographic showing the process of CAR T Therapy.

Source: ACCESS Health International

Using T cells to combat cancer is used in more than just checkpoint inhibitor therapies and is likely the way of the future of cancer treatments. CAR T therapies are a prime example of recent and future innovations in cancer care. Like checkpoint inhibitors, CAR T weaponizes T cells against cancer; the difference is that CAR T involves genetically modifying our T cells to recognize cancer cells and attack them precisely.

By altering our T cells and equipping them with a novel receptor that targets CD19, a specific antigen found on the surface of both cancerous and non-cancerous B cells, they can effectively combat cancer cells. Once T cells are primed with this new chimeric receptor, scientists stimulate their multiplication. After generating a significant number of these cells, they can be reintroduced into the body to effectively target and eliminate cancer cells.

Since 2017, the FDA has approved a half dozen CAR T therapies for blood cancers. CAR T has become an essential component of modern cancer treatment. Clinical trials have yielded promising results, with many patients enjoying long-term remission after receiving treatment with CAR-Ts.

The innovations in CAR T treatments are still expanding. One example is the CAR T switchblade. The idea is to create an antibody switch that controls CAR T cells. This tunable response may overcome the translational barrier of CAR T-cell therapy. Early studies show these switchable CAR T cells are safer for patients.

The future of cancer therapy likely lies in the coming together of many of these treatment types. The development and refinement of each will allow it to be more effective in singular and combination treatments. Studies are underway to determine how to combine checkpoint inhibitors with CAR T therapies and chemotherapies.

To learn more about CAR T and its applications, watch for the release of a forthcoming book titled: *CAR-T: A New Cure for Cancer, Autoimmune and Inherited Diseases.* This book will cover CAR T applications for cancer, autoimmune diseases, and inherited diseases.

There are [more than 200 types of cells](#) in the human body, and T cells are just one of them.

Using Your Cells to Heal Yourself

There are trillions of cells in the human body, around 37 trillion cells, to be more precise. These microscopic marvels make up all our tissues which make up our organs. Even more compelling is that just one cell holds all the information needed to create a person. It contains all it needs to make eyes, arms, hair, and brain.

Every cell is a tiny universe, encased by a membrane and buzzing with cytoplasm. Cells are incredibly small, with most measured in micrometers (µm), which is one-millionth of a meter. To put this into perspective, the average human hair is about 100 µm in diameter, meaning that cells are about ten times smaller than a single strand. They are also much smaller than the naked eye can see, with most cells around 10-30 µm in diameter. The period at the end of this sentence is about 500 µm in diameter, making cells only a fraction of its size. Despite their small size, cells are vital in keeping our bodies functioning correctly.

Out of the 37 trillion cells that constitute our body, about 330 billion of them are replaced daily, meaning that over 3.8 million cells are created in our body every second of the day.

Cellular medicine seeks to harness this regeneration potential to heal the body. Cell therapy involves replacing diseased or damaged cells with new, healthy ones. It encompasses modulating cell function through direct interaction and eliminating disease-causing cells with the aid of immune cells. That means all the immunotherapies discussed earlier are forms of cell therapies. Moving forward, we can now direct our attention to the regenerative capabilities of these cell therapies.

Currently, cells are used *in vitro* to regenerate specific tissues. The process starts by identifying a source of cells, such as stem cells. Stem cells can differentiate into any cell in the body, given the right circumstance and environment. When a cell becomes specialized, it will significantly change size, shape, metabolic activity, and function.

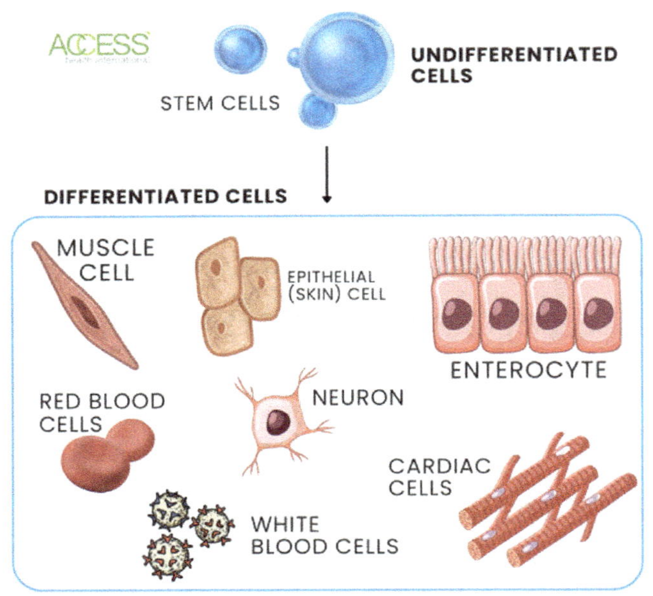

Figure 2. An infographic showing undifferentiated stem cells becoming differentiated cell types.

Source: ACCESS Health International

Cellular medicine relies on various stem cell sources, including embryonic stem cells, induced pluripotent stem cells (iPSCs), and adult stem cells. Induced pluripotent stem cells arise when reprogramming factors, known as the Yamanaka factors, are introduced to a given cell type.

Creating induced pluripotent stem cells involves four essential steps: isolating and culturing donor cells, converting them into stem cells through gene addition, harvesting and culturing them, and finally obtaining the desired induced pluripotent stem cells. Viral vectors are crucial in delivering genetic material into the cells during this process.

Since their discovery in 2006, induced pluripotent stem cells have unveiled a new realm of therapeutics and biomedical research. Stem cells are also noteworthy because they can be made from a patient's body, making them patient-matched or autologous. Since these cells come from the patient, the risk of immune rejection is low.

From a biomedical and drug development perspective, stem cells and induced pluripotent stem cells are significant breakthroughs as they allow for the creation of better disease models. Researchers can make cells with these same mutations by reprogramming cells from patients with genetic diseases. After these cells, they can force them to differentiate into a needed cell type to study targeted therapies' disease mechanisms and effectiveness.

If we can grow cells this way, can we grow whole tissues and organs?

Growing Something New

Organoids, which are structures composed of stem cells that emulate the characteristics of specific tissues or organs, have become a critical tool in biomedical research and regenerative medicine. They provide essential insights into disease processes and drug development, making them a promising avenue for studying disease progression, drug development, and organ development. The organoids are typically created from induced pluripotent stem cells or primary tissue samples through self-assembly.

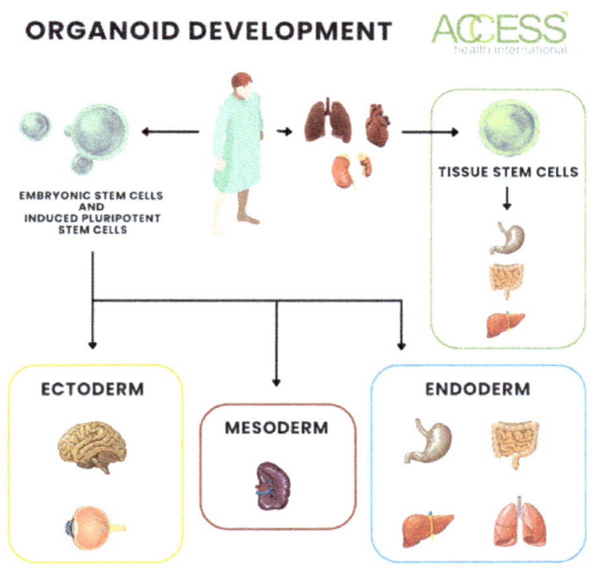

Figure 3. Infographic illustrating the organoid development process.

Source: ACCESS Health International

Compared to traditional 2D culture systems, [organoids offer many advantages](#), such as better recapitulation of *in vivo* physiology and the ability to model diseases more accurately and efficiently. For instance, studies have shown organoids to be highly effective in modeling Alzheimer's and Parkinson's diseases, where the cells used for organoid creation mimic the degeneration of neurons. This capability is essential for understanding how these diseases develop and how to counter their impacts.

Researchers have made significant progress in producing organoids that closely resemble various organs, including the brain, kidney, lung, intestine, stomach, liver, pancreas, thyroid, and retina. These advances have opened up new possibilities for studying and treating organ-specific diseases. You can read more about organoids and their medical uses on my website and in my Forbes column.

One notable breakthrough is the groundbreaking method developed by researchers at Eindhoven University of Technology for growing functional kidney organoids. They have successfully grown kidney organoids with functional glomeruli, crucial in waste filtration. This progress is a significant step towards realizing fully functional kidney organoids, which could revolutionize the treatment of kidney-related ailments.

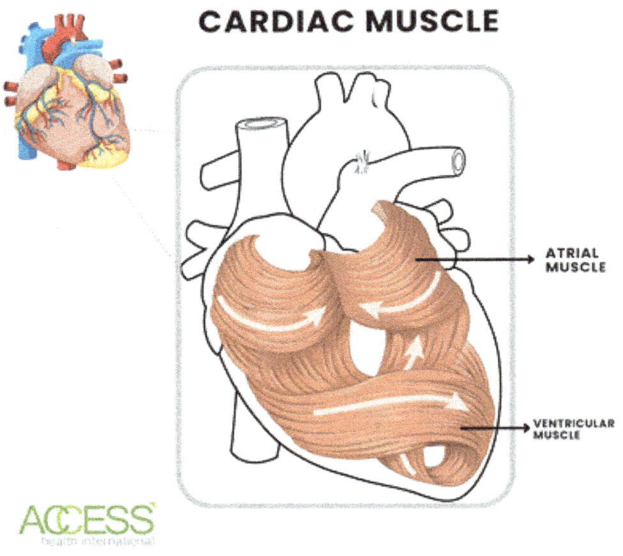

Figure 4. Illustration showing the structure of cardiac muscle..

Source: ACCESS Health International

In [another breakthrough](), scientists from Harvard University have replicated the intricate helical pattern of human hearts at a microscopic scale, resulting in functional heart organoids that demonstrate spontaneous contractions like those observed in a natural heart.

Researchers are also working on the development of eye and brain organoids, which hold immense potential for treating eye-related ailments such as macular degeneration, studying the development and function of the human brain, and modeling neurodevelopmental disorders such as autism and schizophrenia. Collaborative efforts among various research groups are poised to deepen our understanding of these issues and pave the way for innovative treatments and therapies.

Transplantation Innovation

Organ transplants are still a relatively new medical procedure, with the first successful transplant being performed in 1954 when Dr. Joseph Murray conducted a kidney transplant between identical twins. Since then, organ transplants have saved countless lives and improved the quality of life for patients worldwide. Today, they have become a standard treatment for end-stage organ failure.

Transplants involve surgical transplantation from one person to another. But what if we could transplant organs, tissues, or cells from one species to another? This question has been on medical scientists' minds since the early 1900s. The first attempted xenotransplantations used chimpanzee kidneys.

One breakthrough in organ transplantation has been the development of immunosuppressive drugs, which help prevent the body from rejecting a transplanted organ. These drugs have allowed people to receive transplants from donors who are not a perfect match, greatly expanding the pool of potential donors. In recent years, there have also been advances in tissue engineering, allowing scientists to grow organs in the lab using a patient's cells.

Despite these advances, organ transplantation still faces many challenges. One major hurdle is the shortage of donor organs, which has led to long waiting lists and difficult decisions about who should receive a transplant.

More recently, the focus has been on pig-to-human transplants. Xenotransplantation of this kind is feasible due to genetic modifications in pigs, resulting in the production of organs that exhibit greater compatibility with humans. Modified pig organs have been used in successful kidney and heart transplants.

New York University (NYU) researchers conducted one such successful xenotransplantation procedure. They transplanted a genetically modified pig kidney into a deceased patient. The kidney functioned normally and avoided rejection, marking a significant step towards addressing the global organ shortage.

Some challenges arise during organ transplantation, such as endogenous retroviruses in pig DNA and the inability of the immune system to accept foreign animal tissues. However, advances in gene editing and biomaterials offer hope for overcoming these obstacles and transforming the field of organ transplantation.

Ongoing research and development in xenotransplantation aim to refine the techniques and improve the long-term success of pig-to-human organ transplants. Continued research and advances in this field may lead to more successful and widespread use of xenotransplantation. Read more about transplants and xenotransplantation on [my website](#)and in [my Forbes column.](#)

The Future is Now

Healthcare innovations are not stopping at CAR T and xenotransplants. Some of the newest discoveries also include brain-machine interfaces or BMIs. These BMIs allow direct communication between the brain and an external device. Usually, these external devices are computers or prosthetics.

As a medical breakthrough, these interfaces have the potential to give function and movement back to those with neurological disorders or injuries. For example, there are currently [brain-machine interfaces being developed that help individuals with paralysis to communicate](#) via computer. The device records brain activity from the scalp's surface or via implanted sensor. From there, computers can process and interpret the activity to generate signals that control various tools like computers, prosthetics, and wheelchairs.

Bioimplants, like the one mentioned in the prior paragraph, repair the physical integrity of any damaged biological structures. The demand for these novel implants is driven primarily by the ever-rising prevalence of chronic diseases like cardiovascular disease. For example, [the PeriCord implant](#) promotes damaged tissue revascularization and improves cardiac function for patients who have had heart attacks and thus have infarcted heart tissues.

Biomechanical prosthetics offer unprecedented possibilities for patients, expanding once unthinkable horizons. Among those profoundly affected are individuals who have undergone limb amputation due to complications from diabetic foot ulceration. The National Diabetes Foot Care Audit from England found [nearly 60,000 diabetic patients suffer from these ulcerations](#).

Given the vast number of individuals who stand to gain from these robotic prosthetics, it's hardly astonishing that the worldwide market is projected to surge by more than $1.8 billion within the coming decade. Though BMIs are just one example of how technology is revolutionizing regenerative medicine. Another exciting area of innovation is using the vagus nerve to treat various conditions, including inflammatory diseases, depression, and epilepsy.

The vagus nerve is the longest cranial nerve in the body and is vital in regulating various bodily functions, including digestion, heart rate, breathing, and immune function. It is a crucial communication link connecting the brain with the entire body. The vagus nerve performs four primary functions: sensory, special sensory, motor, and parasympathetic. When activated, this extraordinary nerve orchestrates a reduction in heart rate, a decrease in blood pressure, and an enhancement in digestion.

By stimulating this nerve, we can treat various diseases and disorders. This approach involves using electrical impulses to stimulate the left vagus nerve. Healthcare providers implant a small device in the chest under the skin, and a wire runs under the skin connecting the device and the nerve. The device sends mild, painless electrical signals to the brain through the left vagus nerve, calming down irregular electrical activity. Stimulating the vagus nerve has shown promising results in treating various conditions, including depression, epilepsy, and gastroparesis.

Vagal nerve stimulation has been [FDA-approved](#) for treating epilepsy, depression, anxiety disorders, and cluster headaches. It involves implanting a small device under the chest that sends electrical impulses to the vagus nerve. [Recent studies published in *Proceedings of the National Academy of Sciences (PNAS)*](#) suggest it may be effective for chronic inflammatory conditions like rheumatoid arthritis and Crohn's disease. While questions remain, further clinical trials are needed to explore its potential as a treatment option. Understanding the vagus nerve's various roles brings us closer to new therapeutic targets.

Regenerative medicine has seen numerous innovations; the ones discussed are just the beginning. Cellular medicine was the starting point, but much more must be discovered and developed. Medical advances are brimming with exciting possibilities as scientists delve into cutting-edge innovations like biomaterials, organoids, gene therapies, brain-machine interfaces, and more.

These advances hold immense potential to transform medical diagnoses and treatments, making healthcare more efficient and widely accessible. As we witness this progress, our perception of healthcare is shifting, and the future holds even more promise.

References

BCC Research. (2021). Global DNA Repair Drugs Market Size, Share & Trends Analysis Report By Application (Breast Cancer, NSCLC, Ovarian Cancer), By Region, And Segment Forecasts, 2021 – 2028. Retrieved from

https://www.bccresearch.com/market-

research/biotechnology/dna-repair-
drugs-market.html

Cooper, D. K. (2012). A brief history of
cross-species organ transplantation.
Proceedings (Baylor University.
Medical Center), 25(1), 49–57.
https://doi.org/10.1080/08998280.2012
.11928783

Data Bridge Market Research. (2022).
Global Nucleic Acid Based Drugs
Market: Industry Trends and Forecast
to 2028. Retrieved from
https://www.databridgemarketresearch.
com/reports/global-nucleic-acid-based-
drugs-market

Dobosz, P., & Dzieciątkowski, T. (2019). The Intriguing History of Cancer Immunotherapy. Frontiers in immunology, 10, 2965. https://doi.org/10.3389/fimmu.2019.02965

Electrocore. (n.d.). FDA releases gammaCore, the first non-invasive vagus nerve stimulation therapy applied at the neck for acute treatment of pain associated with episodic cluster headache in adult patients. Retrieved November 7, 2022, from https://www.electrocore.com/news/fda-releases-gamm

Fus-Kujawa, A., Mendrek, B., Trybus, A., Bajdak-Rusinek, K., Stepien, K. L., & Sieron, A. L. (2021). Potential of Induced Pluripotent Stem Cells for Use in Gene Therapy: History, Molecular Bases, and Medical Perspectives. Biomolecules, 11(5), 699. https://doi.org/10.3390/biom11050699

GM Insights. (2022). Antibody Therapy Market Share 2022 Growth, Statistics, Trends, Forecast Report. Retrieved from https://www.gminsights.com/industry-analysis/antibody-therapy-market

GM Insights. (2022). Monoclonal Antibodies Market Size By Type, By Application, By Indication, By Distribution Channel Analysis Report, Regional Outlook, Growth Potential, Price Trends, Competitive Market Share & Forecast, 2022 – 2026. Retrieved from https://www.gminsights.com/industry-analysis/monoclonal-antibodies-market

GM Insights. (n.d.). Robotic prosthetics market share analysis & trends to 2025. Retrieved November 7, 2022, from

https://www.gminsights.com/industry-
analysis/robotic-prosthetics-market

Healthcare Conferences UK. (n.d.).
Diabetic Foot Problems: prevention
and management. Retrieved
November 7, 2022, from
https://www.healthcareconferencesuk.
co.uk/conferences-
masterclasses/diabetic-foot-
problems#:~:text=%E2%80%9CEach
%20year%20in%20England%20appro
ximately,death%20rate%20from%20ca
rdiac%20events.%E2%80%9D

Jin, S., Sun, Y., Liang, X. et al. Emerging
new therapeutic antibody derivatives

for cancer treatment. Sig Transduct Target Ther 7, 39 (2022). https://doi.org/10.1038/s41392-021-00868-x

Karolinska Institutet. (n.d.). Timeline - NK cells. Retrieved from https://ki.se/en/research/timeline-nk-cells

Khan YS, Farhana A. Histology, Cell. [Updated 2023 May 1]. In: StatPearls [Internet]. Treasure Island (FL): StatPearls Publishing; 2023 Jan-. Available from:

https://www.ncbi.nlm.nih.gov/books/N
BK554382/

Lehmann, R., Lee, C. M., Shugart, E.
     C., Benedetti, M., Charo, R. A.,
     Gartner, Z., Hogan, B., Knoblich, J.,
     Nelson, C. M., & Wilson, K. M.
     (2019). Human organoids: a new
     dimension in cell biology. Molecular
     biology of the cell, 30(10), 1129–1137.
     https://doi.org/10.1091/mbc.E19-03-
     0135

M Monguio-Tortajada and others,
     Immunomodulatory effect of first-in-
     human PeriCord cardiac bioimplant:
     preliminary data of the PERISCOPE

clinical trial, European Heart Journal, Volume 43, Issue Supplement_2, October 2022, ehac544.1220, https://doi.org/10.1093/eurheartj/ehac 544.1220

Montgomery, R. A., Stern, J. M., Lonze, B. E., Tatapudi, V. S., Mangiola, M., Wu, M., Weldon, E., Lawson, N., Deterville, C., Dieter, R. A., Sullivan, B., Boulton, G., Parent, B., Piper, G., Sommer, P., Cawthon, S., Duggan, E., Ayares, D., Dandro, A., Fazio-Kroll, A., Stewart, Z. A. (2022). Results of two cases of pig-to-human kidney xenotransplantation. The New

England journal of medicine, 386(20), 1889–1898. https://doi.org/10.1056/NEJMoa2120238

Myers, J.A., Miller, J.S. Exploring the NK cell platform for cancer immunotherapy. Nat Rev Clin Oncol 18, 85–100 (2021). https://doi.org/10.1038/s41571-020-0426-7

National Cancer Institute. (n.d.). CAR T Cells: Engineering Patients' Immune Cells to Treat Their Cancers. Retrieved from https://www.cancer.gov/about-cancer/treatment/research/car-t-cells

Pandarinath, C., Nuyujukian, P., Blabe, C. H., Sorice, B. L., Saab, J., Willett, F. R., Hochberg, L. R., Shenoy, K. V., & Henderson, J. M. (2017). High performance communication by people with paralysis using an intracortical brain-computer interface. eLife, 6, e18554. https://doi.org/10.7554/eLife.18554

Physiopedia. (n.d.). Vagus Nerve. Retrieved November 7, 2022, from https://www.physio-pedia.com/Vagus_Nerve

Science. (2022, November 4). Scientists report first successful human trial of gene editing with CRISPR in patients with inherited blindness. Science | AAAS. https://www.science.org/doi/10.1126/science.abl6395

Sender, R., Milo, R. (2021). The distribution of cellular turnover in the human body. Nat Med 27, 45-48.

https://doi.org/10.1038/s41591-020-01182-9

Stepanov, A., Kalinin, R., Shipunova, V., Zhang, D., Xie, J., Rubtsov, Y., Ukrainskaya, V., Schulga, A., Konovalova, E., Volkov, D., Yaroshevich, I., Moysenovich, A., Belogurov, A., Zhang, H., Telegin, G., Chernov, A., Maschan, M., Terekhov, S., Wu, P., Deyev, S., Lerner, R., Gabibov, A., & Altman, S. (2022). Switchable targeting of solid tumors by BsCAR T cells. Proceedings of the National Academy of Sciences, 119 (46) e2210562119.

Wang, V., Gauthier, M., Decot, V., Reppel, L., & Bensoussan, D. (2023). Systematic Review on CAR-T Cell Clinical Trials Up to 2022: Academic Center Input. Cancers, 15(4), 1003. https://doi.org/10.3390/cancers150410 03.

9 798986 133454